The Outer Planets

by Christina Leaf
Illustrated by Natalya Karpova

BELLWETHER MEDIA
MINNEAPOLIS, MN

Blastoff! Missions takes you on a learning adventure! Colorful illustrations and exciting narratives highlight cool facts about our world and beyond. Read the mission goals and follow the narrative to gain knowledge, build reading skills, and have fun!

Traditional Nonfiction

Narrative Nonfiction

Blastoff! Universe

MISSION GOALS

> FIND YOUR SIGHT WORDS IN THE BOOK.

> LEARN ABOUT THE OUTER PLANETS OF THE SOLAR SYSTEM.

> CHOOSE A PLANET YOU WOULD LIKE TO LEARN MORE ABOUT.

This edition first published in 2023 by Bellwether Media, Inc.

No part of this publication may be reproduced in whole or in part without written permission of the publisher. For information regarding permission, write to Bellwether Media, Inc., Attention: Permissions Department, 6012 Blue Circle Drive, Minnetonka, MN 55343.

Library of Congress Cataloging-in-Publication Data

Names: Leaf, Christina, author.
Title: The outer planets / by Christina Leaf.
Description: Minneapolis, MN : Bellwether Media, 2023. | Series: Blastoff! missions. Journey into space | Includes bibliographical references and index. | Audience: Ages 5-8 | Audience: Grades 2-3 |
Summary: "Vibrant illustrations accompany information about the outer planets of the solar system. The narrative nonfiction text is intended for students in kindergarten through third grade"-- Provided by publisher.
Identifiers: LCCN 2022006870 (print) | LCCN 2022006871 (ebook) | ISBN 9781644876572 (library binding) | ISBN 9781648348419 (paperback) | ISBN 9781648347030 (ebook)
Subjects: LCSH: Outer planets--Juvenile literature.
Classification: LCC QB659 .L43 2023 (print) | LCC QB659 (ebook) | DDC 523.4--dc23/eng20220422
LC record available at https://lccn.loc.gov/2022006870
LC ebook record available at https://lccn.loc.gov/2022006871

Text copyright © 2023 by Bellwether Media, Inc. BLASTOFF! MISSIONS and associated logos are trademarks and/or registered trademarks of Bellwether Media, Inc.

Editor: Betsy Rathburn   Designer: Jeffrey Kollock

Printed in the United States of America, North Mankato, MN.

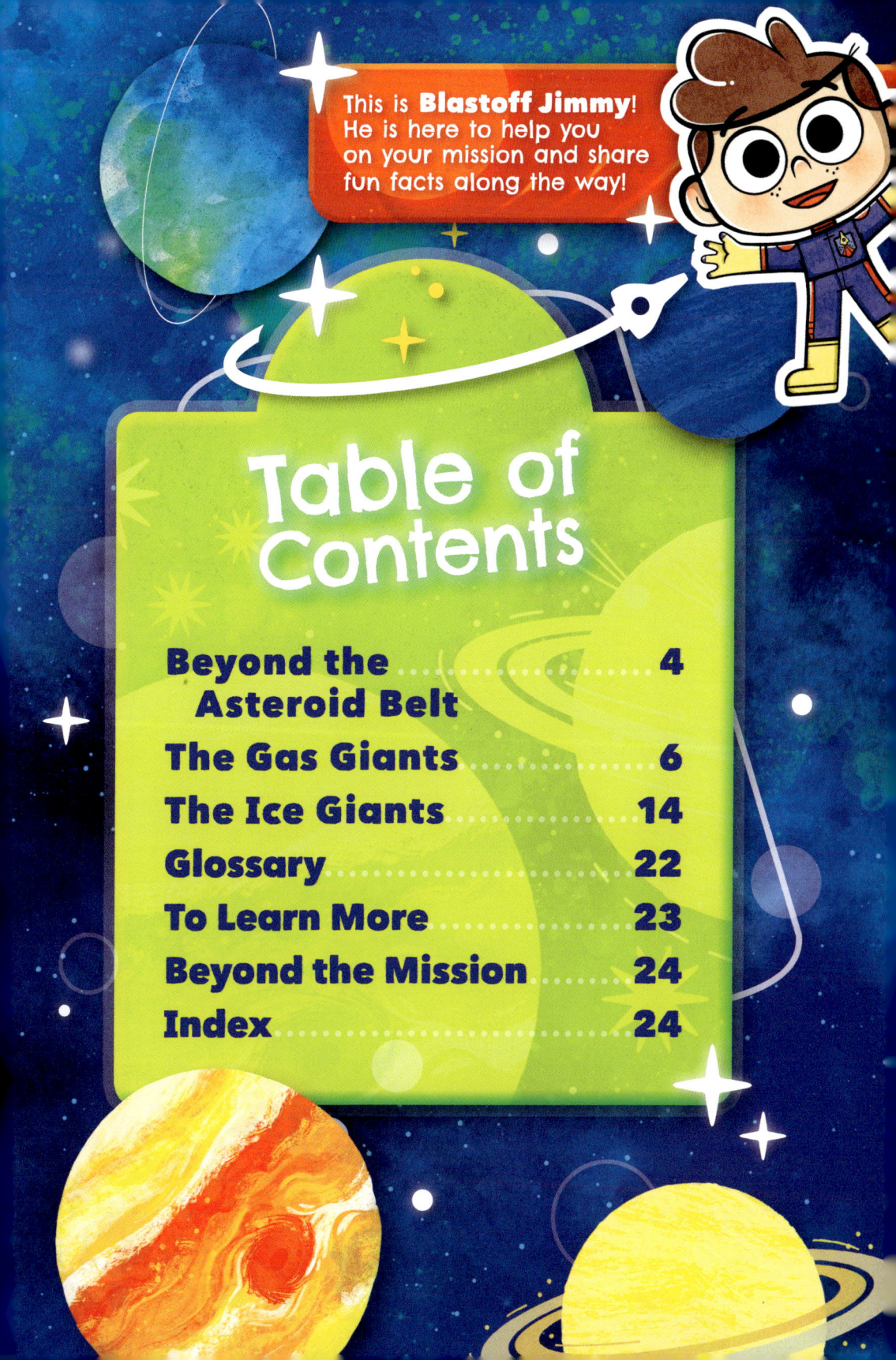

This is **Blastoff Jimmy**! He is here to help you on your mission and share fun facts along the way!

Table of Contents

Beyond the Asteroid Belt	4
The Gas Giants	6
The Ice Giants	14
Glossary	22
To Learn More	23
Beyond the Mission	24
Index	24

You excitedly take your seat. Today your class will learn about the outer planets. They lie beyond the **asteroid belt**. Your imagination takes off as class begins. Here we go!

The Gas Giants

Jupiter

Your first stop is Jupiter, the largest planet in the **solar system**. Colorful clouds form stripes across this **gas giant**. The Great Red Spot swirls in the lower half.

JIMMY SAYS

The Great Red Spot is a big storm. It is bigger than Earth!

Great Red Spot

Ganymede

You try to count Jupiter's moons, but you lose count after 75. You give your spaceship an extra boost to exit Jupiter's powerful **gravity**.

You gasp as you reach Saturn, the sixth planet from the Sun. Its wide **rings** are spectacular! A closer look shows that the rings are chunks of ice and rock.

Saturn

You weave your spaceship around Saturn's moons. There are more than 80!

As you pass Titan, the largest, you look closely. Scientists say the orange moon could hold life!

The Ice Giants

You begin to shiver as you speed toward the **ice giants**. The Sun's warmth is far away.

Pale blue Uranus comes into view. You count 27 moons surrounding the seventh planet.

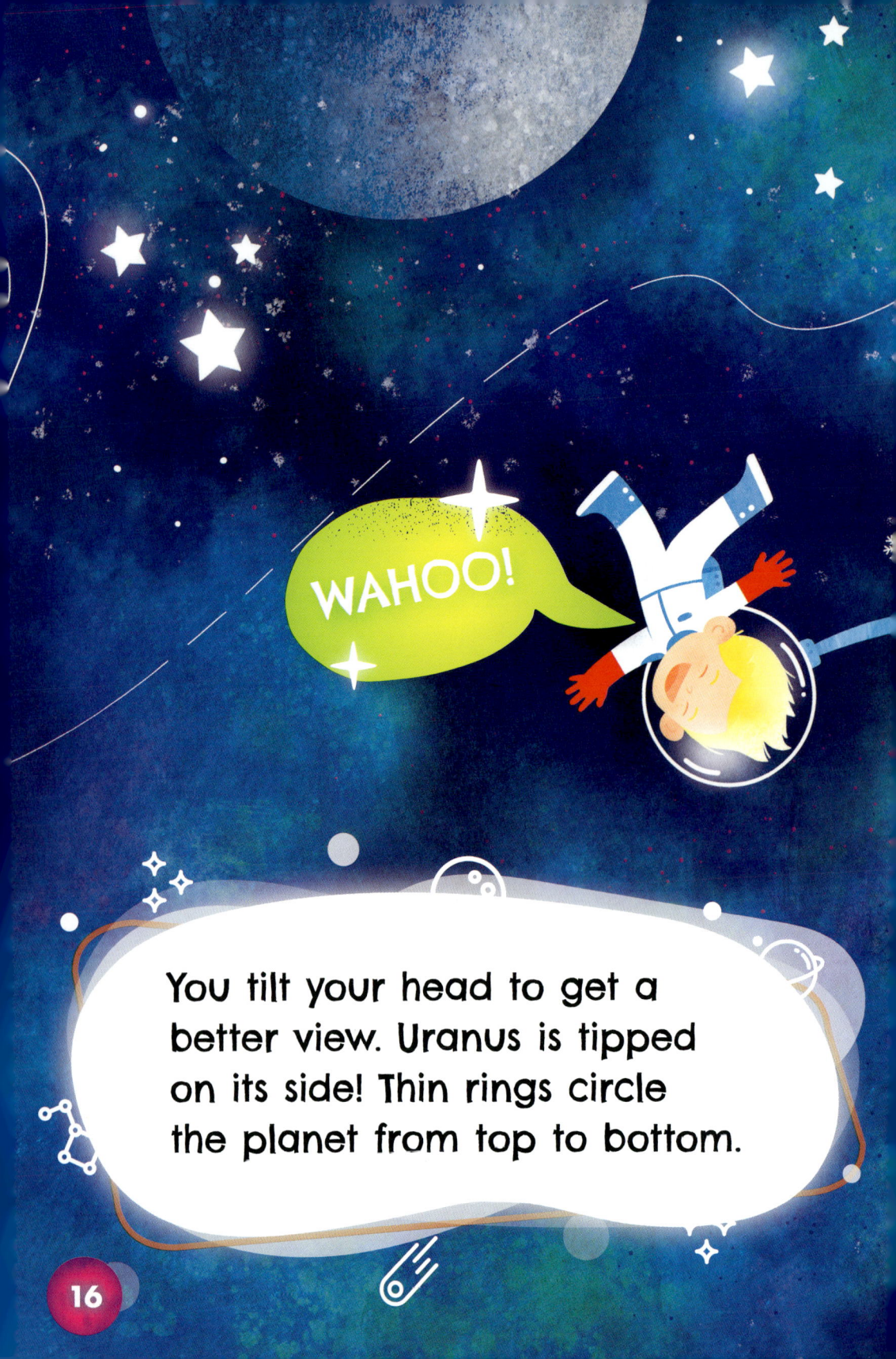

You tilt your head to get a better view. Uranus is tipped on its side! Thin rings circle the planet from top to bottom.

Triton

Neptune

You head to the other ice giant. Neptune's bright blue color hints at the icy winds that whip across the planet. Brrr!

You turn away from Neptune and its moons. From here, the Sun is just a speck. You miss its light and warmth. It is time to come back to Earth. What a journey to the outer planets!

The Outer Planets

Jupiter

Planet Size Rank: 1

Moons: more than 75

Saturn

Planet Size Rank: 2

Moons: more than 80

Uranus

Planet Size Rank: 3

Moons: 27

Neptune

Planet Size Rank: 4

Moons: 14

Glossary

asteroid belt—a part of the solar system between Mars and Jupiter where more than one million asteroids are found

gas giant—a large planet made out of gases that does not have a solid surface; Jupiter and Saturn are gas giants in our solar system.

gravity—a force that pulls objects toward each other

ice giants—large planets that are made of heavier materials than gas giants; Uranus and Neptune are ice giants in our solar system.

rings—bands of small objects like rocks and ice that circle around a planet

solar system—the group of planets, moons, asteroids, and other bodies that circle around the Sun

To Learn More

AT THE LIBRARY

Betts, Bruce. *Super Cool Space Facts: A Fun, Fact-filled Space Book for Kids*. Emeryville, Calif.: Rockridge Press, 2019.

Foxe, Steve. *Neptune*. North Mankato, Minn.: Capstone, 2021.

Leaf, Christina. *The Inner Planets*. Minneapolis, Minn.: Bellwether Media, 2023.

ON THE WEB

FACTSURFER

Factsurfer.com gives you a safe, fun way to find more information.

1. Go to www.factsurfer.com.

2. Enter "outer planets" into the search box and click.

3. Select your book cover to see a list of related content.

BEYOND THE MISSION

> WHICH PLANET IN THIS BOOK WOULD YOU LIKE TO VISIT? WHY?

> DRAW A PICTURE OF WHAT YOU THINK SOMETHING LIVING ON TITAN WOULD LOOK LIKE!

> IF YOU COULD NAME ONE OF JUPITER'S MOONS, WHAT NAME WOULD YOU GIVE IT?

Index

asteroid belt, 5
clouds, 6
Earth, 7, 20
gas giant, 6, 10
gravity, 9
Great Red Spot, 6, 7
ice, 10, 14, 18
ice giants, 14, 18
Jupiter, 6, 8, 9
life, 13
moons, 9, 13, 14, 20
Neptune, 18, 19, 20
rings, 10, 11, 16
rock, 10

Saturn, 10, 12, 13
size, 6, 7, 13, 19
solar system, 6, 19
Sun, 10, 14, 20
Titan, 12, 13
Triton, 18, 19
Uranus, 14, 15, 16
winds, 18